A Merrow Monograph

ACOUSTICAL HOLOGRAPHY

Merrow Monographs

General Editor; J. Gordon Cook, B.SC., PH.D., F.R.I.C.

Concise, authoritative reviews of technical topics

Practical Science Series

ACOUSTICAL HOLOGRAPHY

E. E. Aldridge, B.Sc.(Eng.), C.Eng., M.I.E.E.

Atomic Energy Research Establishment,
Harwell, Didcot, Berkshire, England

MERROW

Merrow Publishing Co. Ltd.
276, Hempstead Road
Watford   Herts   England

ISBN 0 900 54116 4 Clothbound
ISBN 0 900 54117 2 Paperback

Printed in Great Britain at the Pitman Press, Bath

CONTENTS

1

INTRODUCTION

In the late nineteen forties, D. Gabor (1949) demonstrated that it was possible to record a true three dimensional image on a two dimensional photographic plate. He called the process holography. The secret of holography lay in so developing the photographic plate that the phase associated with the image wavefront was cancelled with that of a wavefront corresponding to a spurious image. When the plate was developed and viewed as a transparency in coherent light, the two images were separated visually.

As Gabor showed, holography was not restricted to light but was applicable to all wave systems. As there was no fundamental difference between holograms formed using light and holograms formed using other wave motions, the latter could be turned into optical transparencies and their recorded images viewed optically. This enabled the complexities of image formation in these wave systems to be avoided by using the highly developed optical lens systems.

Acoustical imaging has been under investigation since the nineteen thirties, and general surveys of this work have been given by Rozenberg (1955), Berger and Dickens (1963), Berger (1967 and 1969) and Green (1969), but the advantages of holography in this field do not seem to have been appreciated until the early nineteen sixties. The first to develop acoustical holography appears to have been the Holotron Corporation, a company formed to exploit the developments in holography made at the University of Michigan in the late nineteen fifties and early sixties. Although the work began in

1962 (Worlton, 1968) details were not released generally until 1967.

About 1966–67 a considerable interest in acoustical holography grew in North America, Britain and Japan. During this period a number of short papers were read and published, mainly as letters to the technical press; fields of interest included sonar, medical diagnostics and non-destructive testing. On December 14–15, 1967, the First International Symposium on Acoustical Holography* was held in California, U.S.A. The proceedings of this symposium were published under the joint editorship of Metherell, El-Sum, and Lamore (1969). The bibliographies associated with each paper and the supplementary bibliography added in print, specifically devoted to acoustical holography, cover all the significant papers published up to about the middle of 1968.

Since this time, apart from the possible use of liquid crystals to form acoustical holograms (Lamore, 1969), nothing fundamentally new has emerged. As with all acoustical imaging systems, the problem is to produce a satisfactory acoustical equivalent to the photographic plate. There has been a great deal of practical progress in this respect, but the problem still remains.

Practical advantages of acoustical holography include (a) an image quality much superior to that of the older systems and (b) greater flexibility in use.

* The proceedings of the Second International Symposium on Acoustical Holography have now been published under the same editorship.

2

THEORY OF HOLOGRAMS

In order to obtain a simple physical picture, holograms will be treated from a geometric viewpoint corresponding to thin lens and ray theory. The plane reference beam will be used as an example, as this most closely fits the situation obtaining in acoustical holography. The formal mathematical theory has been given by Gabor (1949) and Leith and Upatnieks (1963).

The construction and image reproduction of a plane reference beam hologram is illustrated in Figs. 1 and 2. No loss of generality results from considering only two dimensions with the object standing on the X axis at a distance x_0 from the holographic plane, and the holographic plane, where the wavefront radiating from the object overlaps the reference beam wavefront, coinciding with the Z axis as shown in Fig. 1.

(a) Hologram Formation

For clarity it will be assumed that the object is illuminated by a plane acoustical wave travelling parallel to the X axis. Now consider a ray, defined as a narrow beam which nevertheless is a sufficient number of wavelengths in cross section, to behave like an infinite plane wave over moderate propagation distances, propagating from the object, at $z = z_0$, to meet the Z axis about z_1 as shown in Fig. 1. Along the Z axis, in the region of z_1, this appears as a wave propagating as:

$$A(z_0) \cos \{\omega t + \phi(z_0) + (2\pi/\lambda) \sin \alpha . z\} \qquad (1)$$

3

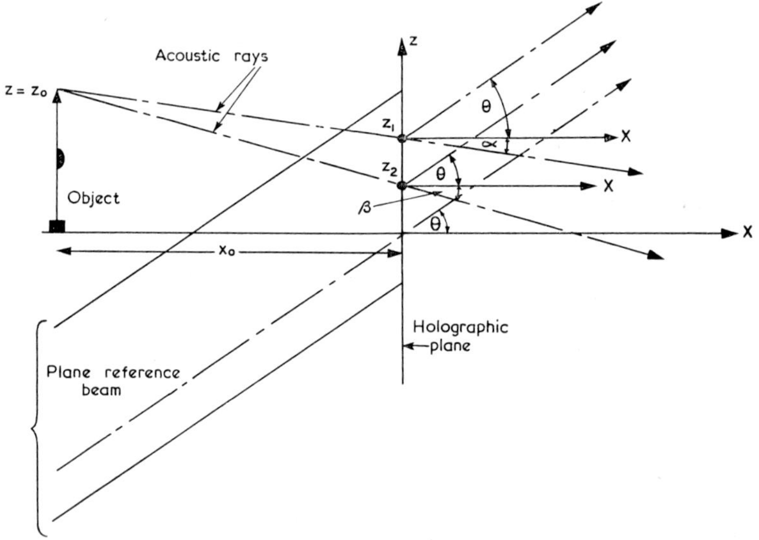

Fig. 1. Formation of a hologram

where $(\omega/2\pi)$ is the acoustic frequency, t time, λ the acoustic wavelength, α the angle the ray makes with the X axis, $A(z_0)$ the amplitude of the ray and $\phi(z_0)$ the phase of the ray relative to some datum. In a similar manner, over the whole of the Z plane, the reference beam propagating at an angle θ to the X axis will appear as a wave propagating as:

$$\cos \{\omega t - (2\pi/\lambda) \sin \theta \cdot z\} \tag{2}$$

where unity amplitude is assumed. Now imagine that in the Z plane these two waves are multiplied together, i.e.:

$$A(z_0) \cos \{\omega t + \phi(z_0) + (2\pi/\lambda) \sin \alpha \cdot z\}$$
$$\times \cos \{\omega t - (2\pi/\lambda) \sin \theta \cdot z\}$$
$$= \tfrac{1}{2} A(z_0) \cos \{(2\pi/\lambda)(\sin \theta + \sin \alpha)z + \phi(z_0)\} \tag{3}$$

where the second harmonic term has been removed. The term that is left is independent of time, and if a constant, B,

4

is added to it to make the sum always positive, then at a position corresponding to z_1, this sum can be recorded as an intensity pattern on a photographic transparency. Similar intensity patterns can be recorded for all the rays from the object, and the photographic transparency so covered with these intensity plots, linearly superimposed, is a hologram.

(b) Hologram Image Reconstruction

The acoustical hologram is now an optical transparency, and the image is viewed by means of coherent light which has a wavelength very much smaller than that of the sound. Again, for clarity, it will be assumed that the illuminating beam is plane and in the plane of the hologram takes the form:

$$\cos \{\omega_1 t - (2\pi/\lambda_1)(\lambda_1/\lambda) \sin \theta . z\} \tag{4}$$

where $(\omega_1/2\pi)$ is the frequency of the light and λ_1 its wavelength. The inclination of this beam to the X axis is

$$\sin^{-1} \{(\lambda_1/\lambda) \sin \theta\} \tag{5}$$

i.e. the angle is reduced approximately by the ratio of the wavelengths when compared to the inclination of the reference beam.

After passing through the transparency, the illuminating beam has its amplitude multiplied by the intensity patterns recorded for the original object rays (this is qualified in Section 3b). For the ray considered, the wavefront emerging from the hologram is obtained by multiplying together expressions (3) and (4), i.e.:

$$\begin{aligned}
\tfrac{1}{2}A(z_0) &\cos \{(2\pi/\lambda)(\sin \theta + \sin \alpha)z + \phi(z_0)\} \\
&\times \cos \{\omega_1 t - (2\pi/\lambda_1)(\lambda_1/\lambda) \sin \theta . z\} \\
&= \tfrac{1}{4}A(z_0) \cos \{\omega_1 t + \phi(z_0) + (2\pi/\lambda_1)(\lambda_1/\lambda) \sin \alpha . z\} \\
&+ \tfrac{1}{4}A(z_0) \cos \{\omega_1 t - \phi(z_0) - (2\pi/\lambda_1) \\
&\qquad\qquad\qquad \times (\lambda_1/\lambda)(2 \sin \theta + \sin \alpha) . z\} \tag{6}
\end{aligned}$$

5

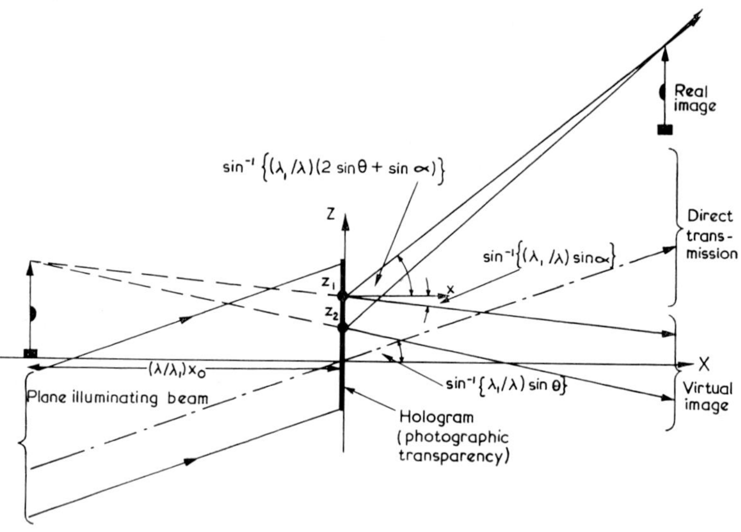

Fig. 2. Formation of an image from a hologram

Comparing the first term on the right-hand side with expression (1), it is seen that it represents a ray having the same amplitude and phase but inclined to the hologram normal at an angle:

$$\sin^{-1}\{(\lambda_1/\lambda)\sin\alpha\} \simeq (\lambda_1/\lambda)\alpha \qquad (7)$$

assuming α is small. Thus the reconstructed ray is identical with the original ray, except that its inclination to the hologram normal is reduced by the ratio of the wavelengths. This is true of all the rays, and by tracing them back to the image, simple geometry shows that they correspond to a distorted virtual image of the object lying at distance $(\lambda/\lambda_1)x_0$ from the hologram.

Using similar arguments, the second term of expression (6) can be shown to correspond to a real image situated at the same distance from the hologram, and in which bumps in the object become dents in this image. The constant term,

B, gives rise to direct transmission of the light through the hologram of amplitude $B \cos \{\omega_1 t - (2\pi/\lambda_1)(\lambda_1/\lambda) \sin \theta \cdot z\}$. As the acoustical reference beam was skew, the two images and the direct transmission are angularly separated, allowing the virtual image to be viewed without interference from the other two. This is illustrated in Fig. 2 using the two rays shown in Fig. 1 and recorded on the hologram in the two regions $z = z_1$ and $z = z_2$.

From the arguments above it is seen that an object has its depth enlarged, in the ratio of wavelengths, in the optical image. This can be ameliorated to some extent by photographically reducing the hologram before viewing. If the reduction is by a factor "m", where "m" is greater than one, the intensity plot for the ray considered in expression (3) becomes:

$$\tfrac{1}{2}A(z_0) \cos \{(2\pi/\lambda)(m \sin \theta + m \sin \alpha)z + \phi(z_0)\} \qquad (8)$$

By following through the previous arguments, it is seen that whilst the image is reduced in size by "m" the distance that the image lies from the hologram is reduced by the square of this factor. Apart from practical considerations, this reduction cannot be taken too far without making the image too small to see conveniently. If, then, a lens system is used to magnify the image, the depth distortion is restored.

3

ACOUSTICAL HOLOGRAMS

Some experimental work on acoustical holograms has been done at frequencies just above audio (Metherell, 1969), but most of the practical work has been at ultrasonic frequencies of 1 MHz to 10 MHz; this covers the frequencies most used in medical diagnostics and non-destructive testing. The wavelengths involved are 1·5 mm to 0·15 mm in water, increasing by a factor between $2\frac{1}{2}$ to 4 for dilatational waves in metals. Both pulsed and continuous ultrasound have been used with water as the medium carrying the sound to the object under view. With continuous ultrasound, it has been found necessary to line the sides and bottom of the tank with sound-absorbent material; this damps out the standing waves which would otherwise form in the water and give rise to poor acoustic images (Holt and Coldrick, 1969).

The optical reproduction requires light of a coherence dependent upon the size of the hologram; generally, the coherence of a laser is not necessary. In practice, however, a laser is always used, as the amount of light diffracted by the hologram into the wanted image is small, and the intensity of light available from a laser makes for convenient viewing. Usually, the image is viewed by means of a television monitor; used in conjunction with the laser, this permits the system to be used in normal light with advantage to the operator. The laser is usually the continuous-wave helium-neon type producing a red light with a wavelength of 0·633 microns.

The propagation of sound is basically a non-linear

phenomenon, and it is only when the amplitudes of vibration are small that it can be considered as a linear wave motion. For this reason, it is convenient to divide the methods of making holograms into two broad classes, (a) *Optical Analogue Systems*, which are the acoustical analogues of the optical systems, and use the non-linear properties of the sound to effect the multiplication, and (b) *Radar Type Systems*, which effect the multiplication externally. The former can be real time systems, but they are relatively insensitive and inflexible when compared to the latter.

(a) Optical Analogue Systems

Representing the wavefront emanating from the object as $A \cos(\omega t + \phi)$ and the reference beam wavefront as $B \cos(\omega t + \gamma)$, both in the plane of the hologram, multiplication is effected by adding the two together and using the acoustic non-linearity as a square law device, i.e.:

$$[A \cos(\omega t + \phi) + B \cos(\omega t + \gamma)]^2$$
$$= B^2 \cos^2(\omega t + \gamma) + 2AB \cos(\omega t + \gamma) \cos(\omega t + \phi)$$
$$+ A^2 \cos^2(\omega t + \phi) \quad (9)$$

This yields for the time independent term:

$$\tfrac{1}{2}B^2 + AB \cos(\gamma - \phi) + \tfrac{1}{2}A^2 \quad (10)$$

If the reference beam illuminates the hologram plane uniformly, the first term in this expression is a constant. The next term gives rise to the holographic action; in conjunction with the first term, if it is large enough, it can be regarded as the hologram proper. The last term is the object wavefront multiplied by itself, and it will give rise to a messy central image in the optical reproduction.

Liquid Surface Levitation

A number of systems have been proposed for exploiting this effect. As yet, the only one which has proved of practical value has been that of liquid surface levitation; in the past, this has been used to form ultrasonic images directly

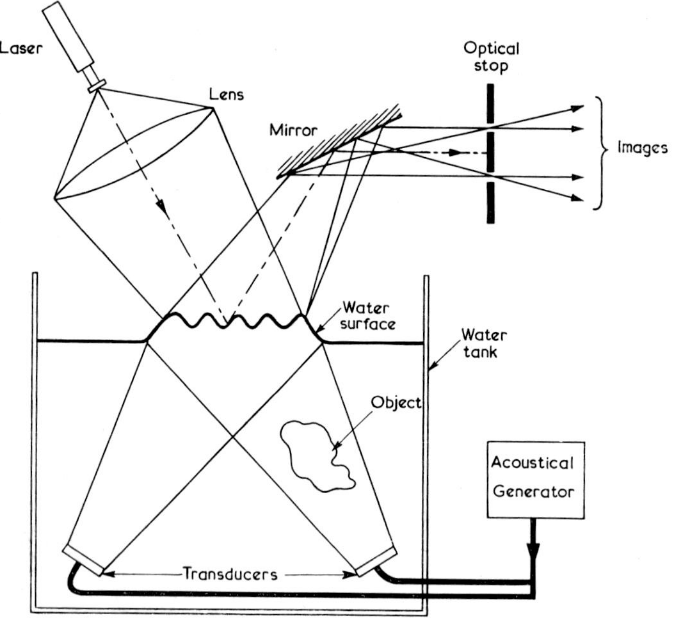

Fig. 3. Holographic system using water surface levitation

(Schuster, 1959). A travelling acoustic wave in a liquid transports liquid, and if an acoustic beam is directed upwards at the surface of the liquid it will produce a permanent displacement of the surface, across the beam wavefront; this is usually several orders of magnitude greater than the amplitude of the surface vibrations. A scheme which utilises this effect is illustrated in Fig. 3.

Two transducers driven from the same generator are placed in a water tank, and oriented so that their beams overlap at the water surface. In addition to the surface displacement, the interference of the beams produces a static ripple pattern. When an object is placed in the path of one of the beams, the other beam serving as the reference, this ripple pattern is modified by the sound scattered by the object. The resultant static surface displacement is then the acoustical hologram. This is used as an optical phase hologram, and the acoustical images are viewed by means of coherent light reflected from it. In absence of the sound, the water surface is a flat plane, and the light reflected from it is focussed to a single spot. With the sound present, but without the object, the resulting surface ripple causes the central spot to be diffracted into a number of others lying symmetrically about it. With the object present, these spots become diffuse, and the images can be seen by viewing the first order diffraction spots. As shown in the figure the others are removed by optical stops.

This arrangement as it stands has two serious disadvantages. First, it is very sensitive to vibration which, in perturbing the water surface, causes the optical images to swing wildly in space; the hologram as such is not particularly affected. Second, the hologram is the same size as the acoustical aperture which, if it is to be useful, is large in terms of light wavelengths. Because of its size, this places rather stringent coherence requirements on the optical illumination.

The commercial version of this system (Holotron Corporation and Brenden, 1967 and 1969) is shown schematically in Fig. 4. Here, the first problem has been solved by forming the hologram on the surface of a shallow auxiliary tank lying in the surface of the main tank. The bottom of this auxiliary tank is transparent to sound; it permits the passage of the acoustical wavefronts without, however, allowing disturbances in the main tank to affect the hologram surface. The object to be viewed is placed in the main tank, and is imaged by means of an acoustic lens onto the plane of the hologram. This ameliorates the second problem; as the image has already been formed, the coherence required of the optical illumination need be sufficient only to enable the reference beam grating to diffract the light. The image is obtained by looking along the appropriate diffracted beam and focussing the optical system on to the plane of the hologram. The sound intensities required are about 0·1 w/cm² (Worlton, 1968), although the threshold sensitivity is about a hundred times less than this (Berger and Dickens, 1963). The ultrasonic frequencies available are 3, 5, 7 and 9 MHz. According to Worlton (1968), the best results have been obtained by pulsing the transducers for approximately 80 microseconds duration at a pulse repetition rate of about 300 per second, and the technique has been successfully extended to ultrasonic frequencies of 50 MHz.

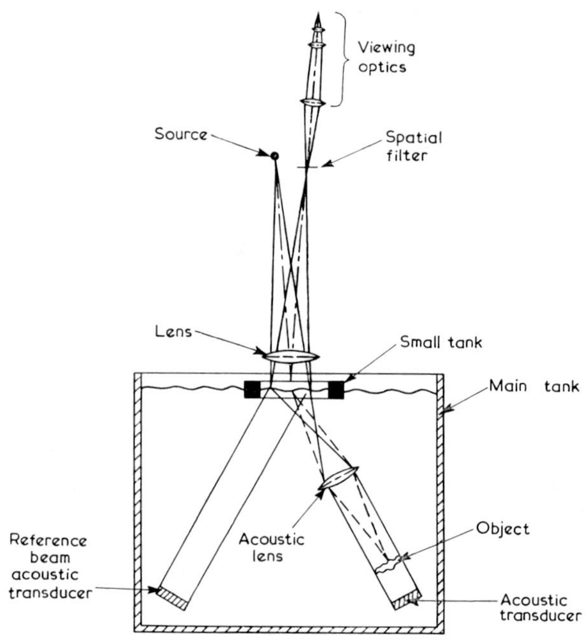

Fig. 4. Commercial holographic system

(b) Radar Type Systems

In this type of system, the wavefront emanating from the object is picked up by a small receiving transducer scanning

12

through it, the scan following some simple geometric form. If the reference beam wavefront is also present, it will be picked up at the same time; if, then, the electrical output of the receiver is passed through a square law detector, the result will be similar to the previous type of system. However, as the receiving transducer is moving through the wavefronts, it is sampling these wavefronts, and these samples appear as sequential signals of the same frequency but different phases. This enables an acoustic reference beam to be replaced with an electrical signal whose phase is varied in sympathy with the movement of the receiving transducer, such that this signal is indistinguishable from the signal which the transducer would produce if it were moving through the wavefront of this beam. Thus, the hologram can be obtained by multiplying together directly, the output of the receiver and this electrical reference using an electronic multiplier, and recording the products sequentially in accordance with the position of the receiving transducer.

Harwell Development

Although there are no commercial versions of this type of system at present, its development is still active. Figure 5 is the schematic of the system developed for the N.D.T. Centre at Harwell, U.K., mainly for the purpose of evaluating the usefulness of acoustical holography in non-destructive testing (Aldridge *et al.*, 1969); it is representative in its essential features.

This figure shows an object in a water bath being illuminated with sound by the send transducer, and the sound scattered by the object being picked up by the receive transducer. There are a number of options; the same transducer can be used for both send and receive (Appendix, Section c) and the object be viewed in direct reflection, or separate transducers can be used and the object viewed in transmission either directly, or indirectly by side illumination. As a transducer for the receiver tends to be insensitive if it is physically small, the focal point of a focussed transducer is

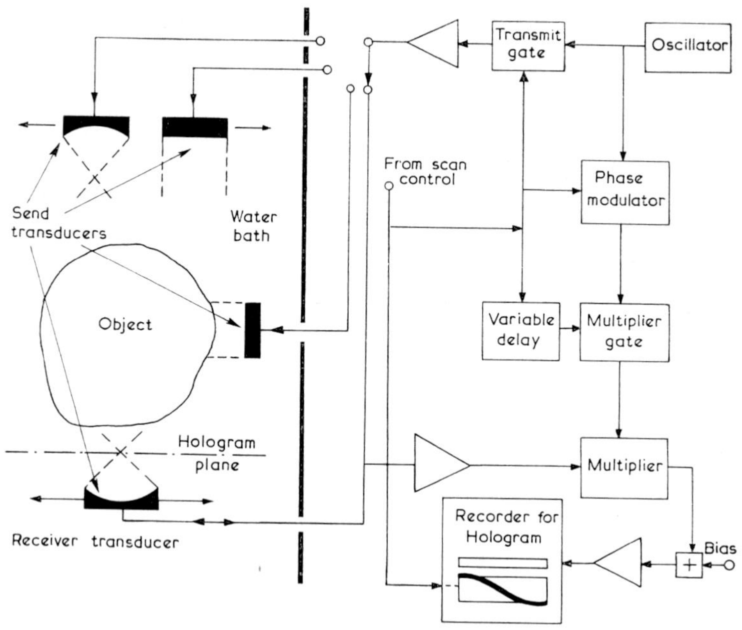

Fig. 5. Scanned holography

used instead (Appendix, Section d). The use of a similar one for the send transducer, as is shown in the figure, will improve the resolution if this is aperture limited (Appendix, Section c). In practice, however, the resolution is limited by the spot size of the receiver (Appendix, Section d), so that the focussing of the send transducer is limited to that which will produce a well-defined narrow beam. An advantage of focussed transducers is that they can be fairly readily made with a spot size to suit the resolution requirements of the job in hand. Where the geometry of the object is favourable, ordinary plane transducers are also used for sending.

Normally, the scan plane of the receive transducer is flat, and the scan is similar to that of a television raster with a fast or line scan in one direction and a slow or frame scan in the perpendicular direction. Where the send and receive transducers face each other, as in direct transmission viewing, then both transducers are scanned together in synchronism (Appendix, Section c). For side illumination, a plane transducer is used with the object in its far zone. The line scan of the receiver is parallel to the beam direction of this transducer, which moves only with the frame scan, i.e. it does not move with the line scan as do the others. For successful holograms, random small-scale irregularities in the scan should not exceed $\lambda/8$ approximately; large-scale irregularities can be tolerated to some extent, as they tend to give rise merely to image distortion.

There are arrangements which make cylindrical scans possible (Appendix, Section e), but these have been little used owing to the small weight of specimen which can be accommodated.

The transducer drive is derived from a crystal oscillator. The output of the oscillator is passed through a transmit gate to produce a short pulse of carrier; this is then transmitted to the send transducer through the output amplifier. The pulse length is usually 1 microsec, but it can be up to several hundreds with a repetition rate lying between 500 p/s to 2500 p/s. The signal from the receive transducer, after passing through the receive amplifier, is passed to a multiplier where it is multiplied by the reference. This latter is obtained by passing the output of the oscillator through a phase modulator and then the multiplier gate. This gate is delayed on the transmit gate by an amount which depends upon which signal pulse is of interest, and serves to produce the range gating facilities for the system. This variable delay, the transmit gate and the phase modulator are synchronized to the scan by means of stepping motors and digital controls. The multiplier is basically a synchronous switch driven by the reference which corresponds to a skew plane beam.

The output of the multiplier has a bias added to it to make it unipolar before it is passed on to the recorder amplifier to produce an intensity plot on a fascimile recorder (Muirhead and Co. Ltd.). The recorder is synchronized to the scan in the same way as before; as its scan is unidirectional, however, its line density is made twice normal, and it records only on every other stroke corresponding to the forward stroke of the receiving transducer. The intensity plot is the ultrasonic hologram; it is photographed and reduced in size to produce the optical transparency.

The development of the transparency is controlled by photographing the grey scale at the same time, and it is arranged to reproduce the tonal contrast of the original. This means that the photographic reduction is limited to about twelve, as the grain size of the emulsion has to be relatively large; also, it means that the density of development is proportional to the signal current fed to the recorder. This latter produces an exponential non-linearity on the light wavefront passing through the transparency, which is compensated by a logarithmic pre-distortion incorporated into the recorder amplifier.

The system was designed for experimental use, using fairly readily available components, to cope with a resolution of one wavelength at 10 MHz in water, about $\frac{1}{6}$ mm, as this was the smallest spot size transducer available. This has necessitated having about 300 scan lines/cm, as determined by a screw thread, and a coverage speed of about $\frac{3}{4}$ sq cm/min. A more sophisticated system would tie the scan density to the resolution required, and this would permit higher coverage speeds at the lower resolutions. With a raster-like scan, operating at constant velocity as here, the maximum speed is set by the total inertia at the motor pinion and the turn-round torque available.

A helium-neon laser is used in the optical reproduction which is illustrated in Fig. 6. The hologram is illuminated by a plane parallel beam, and the transmitted light is brought to a focus by the first lens in its focal plane. As shown, optical stops are placed in this plane in order to remove the diffraction patterns associated with the direct transmission and the

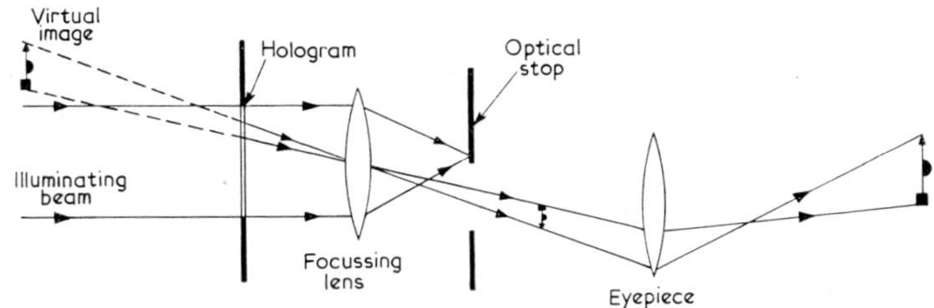

Fig. 6. Reconstructing optics

real image. The virtual image is subsequently re-imaged by an eyepiece. As the angles of the rays diffracted from the hologram are all very small, no attempt has been made to skew the illuminating beam.

A hologram made by scanning has a line structure which, in the image reproduction, will function like a diffraction grating and give rise to beams having angular separations approximately equal to λ/b, referred to the acoustic side, where b is the distance between scan lines. In the hologram, each of these beams will act in the same way as the reference beam and produce an image as illustrated in Fig. 7. If these images are not to overlap and interfere with the main ones, then geometric considerations show that approximately:

$$\lambda/b > 2\theta + \alpha$$

where θ is the angle of the equivalent reference beam and α is the angular aperture of the object. If the scan is not to degrade the resolution of the transducer, then the width between scan lines should not exceed half the spot size of the transducer.

(c) **Performance**

The quality of image possible using simple objects with these systems is illustrated in Fig. 8. As with all the images shown here, this was obtained from the system developed for the N.D.T. Centre, Harwell, U.K.

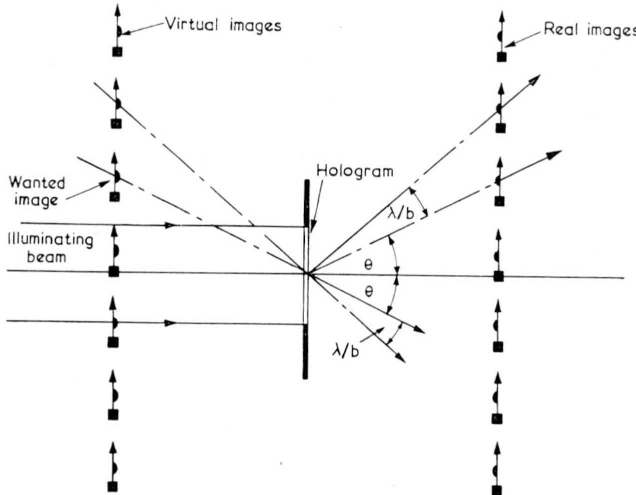

Fig. 7. Multiple imaging of line scan

The object is drawn in Fig. 8(a); it is a piece of PTFE, 2 mm thick, with a window cut in it 15 mm square, having cross-arms 2 mm wide. It was held in position for the scan by means of three 6BA screws. Figure 8(b) shows the image obtained by direct reflection using 10 MHz. Some of the inside edges of the window and one of the outer edges can be seen, indicating that relative to the scanning plane, there was either a slight tilt of the plane of the object or a slight asymmetry in the scanning beam.

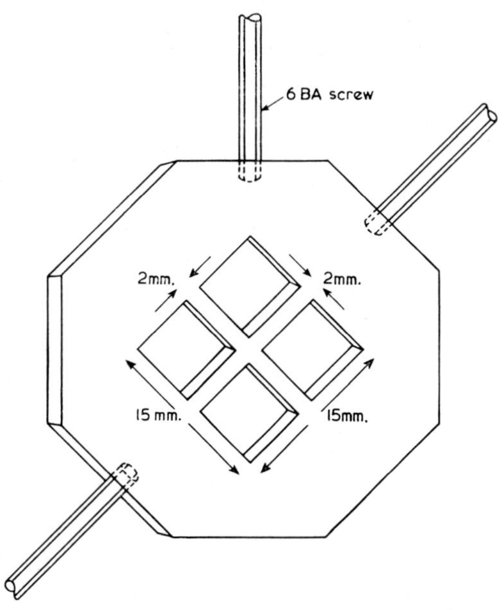

Fig. 8.
(a) Above: Simple object
(b) Right: Image

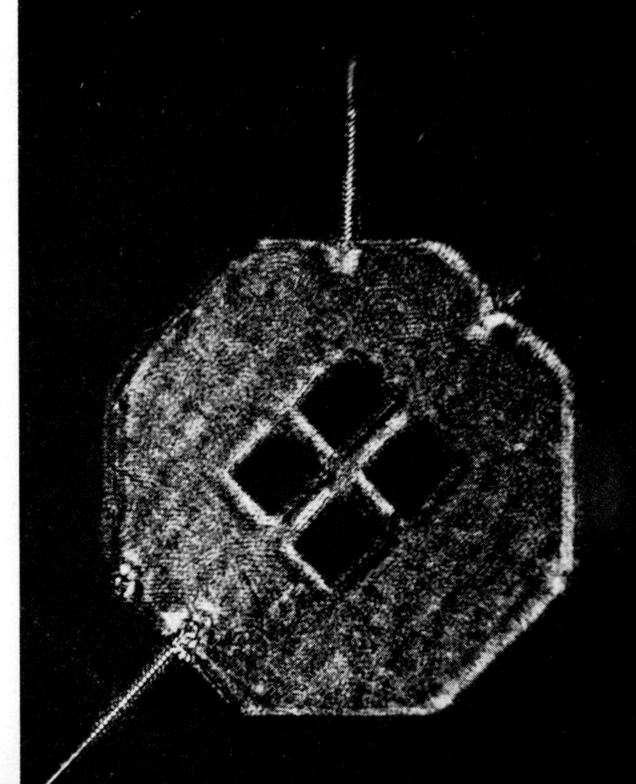

4

PRACTICAL ASPECTS OF ACOUSTICAL HOLOGRAPHY

Comparing sound with light, the wavelengths of the sound frequencies commonly used are much longer than those of light; also, whereas the aperture of optical holograms is large in terms of the wavelength used, the aperture of acoustical holograms is small when expressed in the same terms, although in absolute terms the sizes may be similar. This means that, with the type of object usually shown in optical holograms, the viewer can afford to look at the image through only a small part of the aperture; by choosing where this part lies, the viewer can obtain different views of the image, and a striking 3-D effect. The same thing can be done with acoustical holograms but, because the effective aperture is small and objects of interest are usually comparable in size with the limits of resolution set by this aperture, the resulting loss of resolution is usually unacceptable. This is similar to the situation in optical microscopy, so although depth of focus is present, there is no parallax or 3-D effect.

Sound is invariably generated electronically; this means that it is highly coherent, so that objects viewed by it are seen as spectral images, i.e. cylinders appear as lines, cf. the screws shown in Fig. 8(b). More diffuse images can be produced by introducing a slight degree of spacial phase randomness in the illuminating wavefront. This may be done by transmitting it through some suitable medium or reflecting it from some suitable surface or, in the case of scanned systems, by suitably perturbing the coherence of the

20

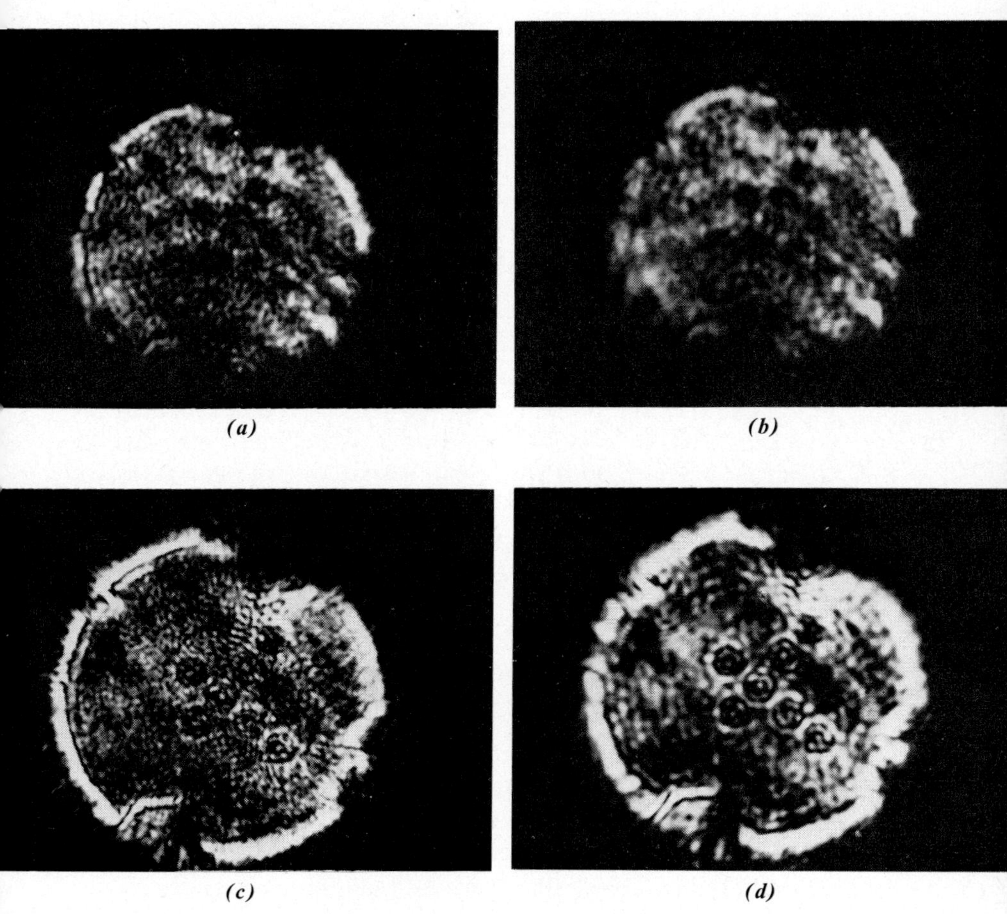

Fig. 9. Image of seven-hole pattern in bottom of stainless steel bar as seen by reflection at 10 MHz through the top of the bar, showing effect of unwanted modes. (a) Transducer spot size 1/6 mm. (b) As for (a) with resolution degraded optically. (c) Transducer spot size 1 mm. (d) As for (c) with resolution degraded optically

generator. Unfortunately, a diffuse image involves a loss of resolution and, in practice, either a loss of intensity if the diffuse wavefront is produced acoustically, or great instrumental complexity if it is produced electronically. Such imaging tends, therefore, to be uncommon.

Effect of Unwanted Modes

In liquids, sound has only one mode of propagation, i.e. compressional, but in solids it has at least two, i.e. basically, compressional and shear. Unfortunately, the imaging process is based on the existence of only one mode; if another is present, the image is confused. As yet, the compressional mode has always been chosen, and in some situations this has meant a lowering of the resolution in order to avoid the unwanted modes. This sort of situation can be a particular difficulty with metallic objects viewed in a water tank. Here, owing to the very large acoustic mismatch, water-metal interface modes can be excited, and these can confuse images of the interior with diffractions from surface discontinuities.

This is illustrated in Fig. 9 which is a series of images, obtained at 10 MHz, of a pattern of seven holes drilled in the bottom of a piece of stainless steel bar. The bar is 5 cm in diameter and $6\frac{1}{2}$ cm long; the holes are conical and were made with the end of a 2 mm-diameter drill on a 4 mm spacing. The view is by direct reflection from the top of the bar looking through the body of the material at the bottom, i.e. the sound has to traverse 13 cm of steel. Figure 9(a) is the image obtained using a transducer with a spot size of approximately $\frac{1}{6}$ mm in water. As the maximum resolution possible in the steel is approximately $\frac{1}{3}$ mm, the scanning transducer is exciting unwanted modes to a considerable degree, resulting in a very poor image as shown in the figure. A slight improvement is obtained by degrading the resolution optically, but as shown in Fig. 9(b) the image is still very poor. Figure 9(c) shows the image obtained using a transducer

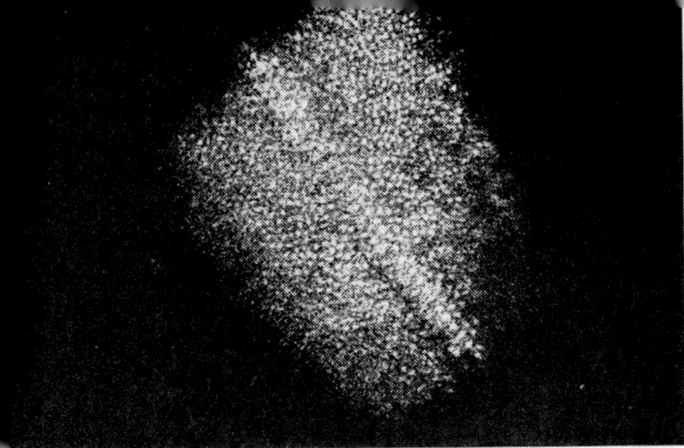

*(a) Left: Top skin of
a piece of plaice*

*(b) Right: View of underly-
ing skeleton where half top
skin has been removed*

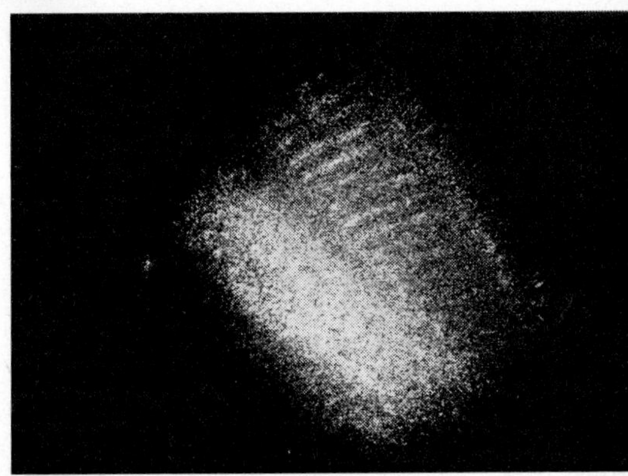

*(c) Left: Skeleton
as   seen   through
under skin*

Fig. 10. Effect of material structure scatter

spot size of about 1 mm; both the outline of the bar and the pattern of the holes are clearly seen. The very fine structure in this image is due probably to the scatter of the sound by the grain structure of the material. If the resolution is degraded optically as for Fig. 9(b), the clouding effect of the fine structure is removed, and the image improves as shown in Fig. 9(d). In the last two figures, the artifacts on the periphery of the bar correlate with draw marks down the side of the bar; this undoubtedly means that the material structure is different in these areas.

Scattering by Material Structure

Often, a comparison is made between X-rays and ultrasound. X-rays are used in transmission and rely on absorption to show up the features of interest in the medium. Ultrasound is not restricted only to transmission; it relies also on diffraction to show up the features of interest. Absorption, when present—as in tissue and bone, and most plastics—is regarded as a nuisance because it limits the depth of penetration. In metals, in practice, absorption is not a problem, but scattering by the material structure, e.g. grain structure, causes difficulties, as it makes the metal behave towards sound as frosted glass behaves towards light. Just as frosted glass confuses the image of any object behind it, so the material structure confuses any image of the material interior, cf. Fig. 9(c) and (d). The problem may be solved, when it is serious, by limiting the ultrasonic frequencies used to those for which the scatter of the material structure is negligible compared to the scatter of the feature of interest.

The effect of scattering of the sound by the material structure is illustrated in Fig. 10; the object is a mid-section of a piece of plaice. The frequency used was 10 MHz and the view is by direct reflection. Figure 10(a) shows the top skin of the fish, and the scales and the underlying backbone are clearly seen. This is almost identical with the optical view. Figure 10(b) is the image obtained when half the top skin is stripped off. Where the skin is missing, the underlying skeleton is clearly seen; where the skin is present, the skeleton is lost in scatter. This skin is like tough tissue paper, and at 10 MHz it scatters the sound considerably. Figure 10(c) shows the complete skeleton when viewed through the soft under-skin of the fish.

5

USES AND POSSIBLE DEVELOPMENTS

In crude terms, it might be said that there are two views of acoustical holography: one is that it complements X-rays and that there should be an acoustical holography machine, like an X-ray machine, working in real time: the other view is that acoustical holography is an extension of the more usual ultrasonic techniques, and like modern radar and sonar, it is a basis for the application of sophisticated information processing. The first point of view leads to the idea of the ultrasonic image as a complement to the X-ray image; the second point of view tends to regard the optical system as a computer which processes the data recorded in the hologram to derive a meaningful pattern, which need not be an image in the usual sense. Considering the two main fields of application, the first point of view is prevalent in medical diagnostics, but both points of view obtain in non-destructive testing.

Medical Diagnostics

In medical diagnostics, the real handicap of acoustical holography, and one that will restrict its use, is that it does not possess any satisfactory equivalent to the photographic plate. Although all the phenomena employed in the past (and some new ones too) for orthodox imaging have been re-examined for possible use in holography, they require too much acoustic power for the patient's well-being or tend to be too inflexible due to the need to accommodate the physical requirements necessary to make the system work. Generally, mechanical systems are not suitable as the total scan time usually has to be much less than one second. It is possible that this problem may be solved in terms of semi-conductor technology, whereby a large piezo-electric element, backed by some integral solid state read out system, is first pulsed to emit sound and then electronically scanned to obtain the field pattern of the reflection. The signal from the read out system would be processed in some manner to yield the image.

Non-Destructive Testing

In non-destructive testing, the situation is a little easier in that for most inspection problems a truly real time system is unnecessary. However, a real time system does enable the inspection of moving parts to be carried out, and it is a field of activity which hitherto has not been explored ultrasonic-ally to any great extent. Because of its complexity of instru-mentation, acoustical holography is not likely to replace the more usual methods, except where inspection is an economic necessity and acoustical holography succeeds where the more usual methods fail. Where the object is small enough, it is probably most convenient to place it in a water bath and use either a fast mechanical scan or a true real time system. The former has the advantage that such scans are in fairly common use and, provided they are stable enough, could be converted to a holographic system merely by changing the backing electronics. Where the object is large, it is most convenient to scan it mechanically using contact probes. In this case, the crucial inspection is often in the early machining stage where the shape of the object is very simple, e.g. a cylinder, and the possibility exists that the scan may be done *in situ* at the same time as the machining. In this sort of situation it would probably be better to record the hologram on tape on the shop floor and only later transfer it to an optical transparency for optical processing. A copy of the optical transparency could be used as a record for purposes such as insurance.

Optical processing (Cutrona, 1964) is rather cumbersome,

3

unless it is just straight imaging, and eventually it will undoubtedly be replaced by digital computer processing. A scheme of operation to avoid excessive computation time would be to use the equivalent of a spherical reference beam in the making of the hologram, and to use range gating and a short pulse so that the hologram was of a thin section of the object; the computer would obtain the image by a Fourier Transform of the hologram (Stroke, 1965) using the Fast Fourier Transform (Cochran *et al.*, 1967).

Acoustical holography could be applied to sonar (Greene and Hildebrand, 1969) and to seismology (Silverman, 1969), but so far very little attempt has been made to do so. The propagation of sound and vibration is very complex in these fields, and as a result holography may well offer no real advantage over the present techniques.

Acknowledgements

To my colleagues Mr. A. B. Clare and Mr. D. A. Shepherd for the photographs of the images shown here.

BIBLIOGRAPHY

E. E. Aldridge, A. B. Clare and D. A. Shepherd, *Proc. Conf. Ind. Ultrasonics*, Loughborough. 245–255. Published I.E.R.E., Bedford Square, London, 1969.

H. Berger (1967), *Acoustical Holography*, Vol. 1, Chap. 2. Plenum Press, New York, 1969.

H. Berger, *Journal Acoustic Soc. Am.*, **45**, (4), 859–867 (1969).

H. Berger and R. E. Dickens, *ANL-6680*, Argonne National Laboratory 1963.

B. B. Brenden (1967), *Acoustical Holography*, Vol. 1, Chap. 4. Plenum Press, New York, 1969.

B. B. Brenden (1969), *Materials Evaluation*, **27** (6), 140–144 (1969).

W. T. Cockran, J. W. Cooley, D. L. Farin, H. D. Helms, R. A. Kaenel, W. W. Lang, G. C. Maling, Jr., D. E. Nelson, C. M. Rader and P. D. Welch, *Proc. I.E.E.E.*, **55**, (10), 1664–1674 (1967).

L. J. Cutrona, *I.E.E.E. Spectrum*, **1**, (10), 101–108 (1964).

D. Gabor, *Royal Soc. Proc.*, **A197**, 454–487 (1949).

D. C. Greene and B. P. Hildebrand, *Ocean Industry*, September, 1969, 43–45 (1969).

P. S. Green, *International Journal of Non-destructive Testing*, **1**, (1), 1–27 (1969).

Holotron Corporation, 1007, Market Street, Wilmington, Delaware 19898.

D. Holt and J. R. Coldrick, *Ultrasonics*, **7**, (4), 240–244 (1969).

L. Lamore, *Ultrasonics*, **7**, (3), 199 (1969).

E. N. Leith and J. Upatnieks, *Journal Optical Society America*, **52**, 1123–1130 (1963).

A. F. Metherell, *Scientific American*, **221**, (4), 36–44 (1969).

A. F. Metherell, H. M. A. El-Sum and L. Lamore, editors, *Acoustical Holography*, Vol. 1. Plenum Press, New York, 1969; Vol. 2. Plenum Press, New York, 1970.

Muirhead and Co., Ltd., Beckenham, Kent.

L. D. Rozenberg, *Soviet Physics-Acoustics*, **1**, 105 (1955).

K. Schuster, *Jenaer Jahrbuch*, 127 (1959).

D. Silverman, *I.E.E.E. Trans. Geoscience Electronics*, **GE-7**, (4), 190–199 (1969).

G. W. Stroke, *App. Phys. Let.*, **6**, 201–203 (1965).

D. C. Worlton, *Battelle Memorial Institute, BNWL-SA-1772* (1968).

APPENDIX—SYNTHETIC APERTURE SCANS

(a) General

For the sake of clarity the analysis is simplified to that of small angle theory. A uniform spherical wave of wavelength λ, and frequency $\omega/2\pi$, propagating outwards from a source at the origin is represented at a radius r by:

$$\frac{1}{r}\,e^{-i\frac{2\pi}{\lambda}r}\,.\,e^{i\omega t}$$

Choosing cartesian axes X and Z, it will be assumed:

$$r^2 = x^2 + z^2$$

i.e. the Y axis is ignored as no loss of generality results. It will also be assumed that the portion of the spherical wavefront which is of interest is that which lies close to the X axis, i.e. all pertinent values of z are much smaller than the values of x. With these assumptions:

$$r \sim x + \frac{z^2}{2x}$$

and

$$\frac{1}{r}\,e^{-i\frac{2\pi}{\lambda}r}\,.\,e^{i\omega t} \sim \frac{1}{x}\,e^{i\left(\omega t - \frac{2\pi}{\lambda}x\right)}\,.\,e^{-i\frac{\pi z^2}{\lambda x}}$$

(b) Plane Illuminating Beam

In Fig. A1 is shown a plane wave, $\exp(i\omega t)$ illuminating an object and the wavefront reflected from the object being

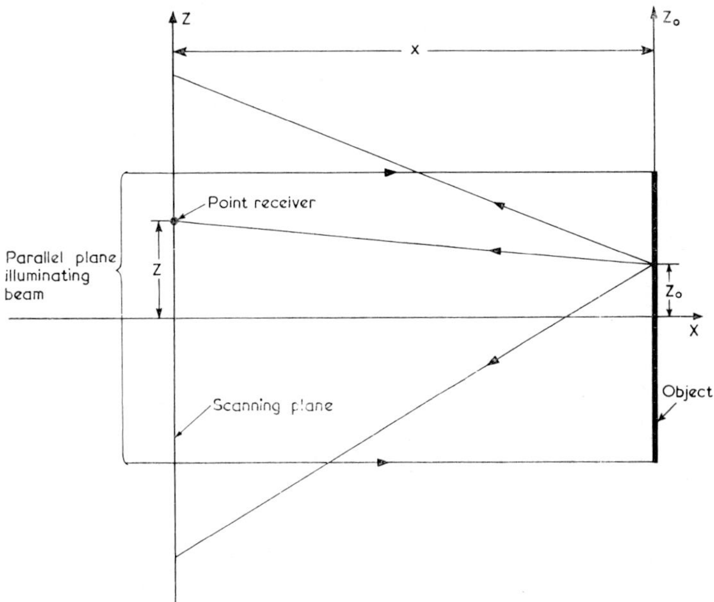

Fig. A.1. Signal received by point receiver with plane wave illumination

sampled by a point receiver. Denoting the reflectivity of the object by $f(z_0)$, then the reflected wavefront is that radiated from a number of point sources of magnitude $f(z_0)dz_0$ in the plane of the object. Hence the signal registered by the point receiver is given by:

$$\frac{1}{x}\,e^{\,i\left(\omega t-\frac{2\pi}{\lambda}x\right)}\int_{z_0}f(z_0)\,e^{-i\frac{\pi}{\lambda x}(z-z_0)^2}\,.\,dz_0 \qquad\qquad \text{A1}$$

where the integration is over the plane of the object. As is shown in the main body of the text, by scanning the point receiver through the wavefront and multiplying its signal by a phase reference to build up a hologram, an image of the object can be obtained.

31

(c) Scanned Source Illumination

In Fig. A2 is shown a spherical wavefront, obtained from a point source, illuminating the object, and the wavefront reflected from the object being sampled by point receiver. In this arrangement the source and the receiver scan the object in synchronism together. To illustrate the salient features of this system it will be assumed that the line joining the source and receiver is perpendicular to the object plane, and that the two scanning planes are parallel to the object plane and to each other, and are at distances x_2 and x_1, respectively, from the object plane.

In the object plane the illumination from the source

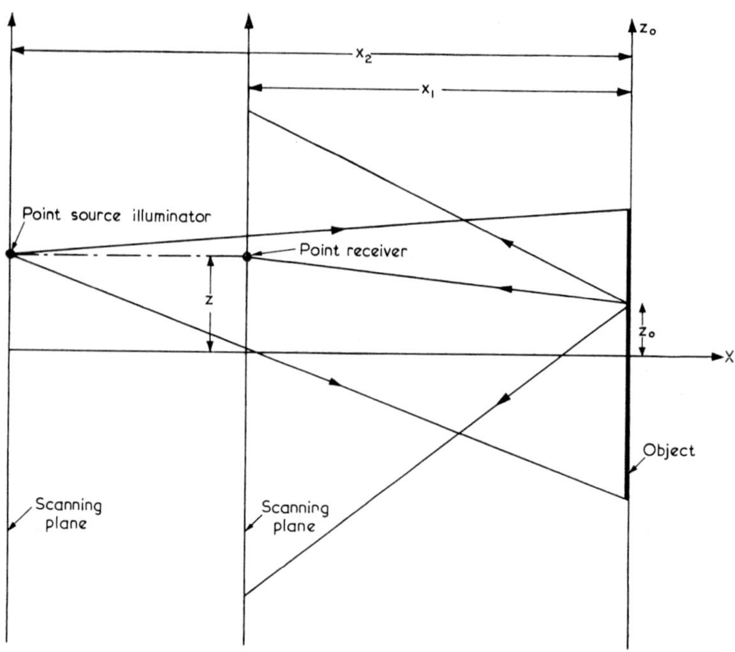

Fig. A.2. Signal received by point receiver with spherical wave illumination

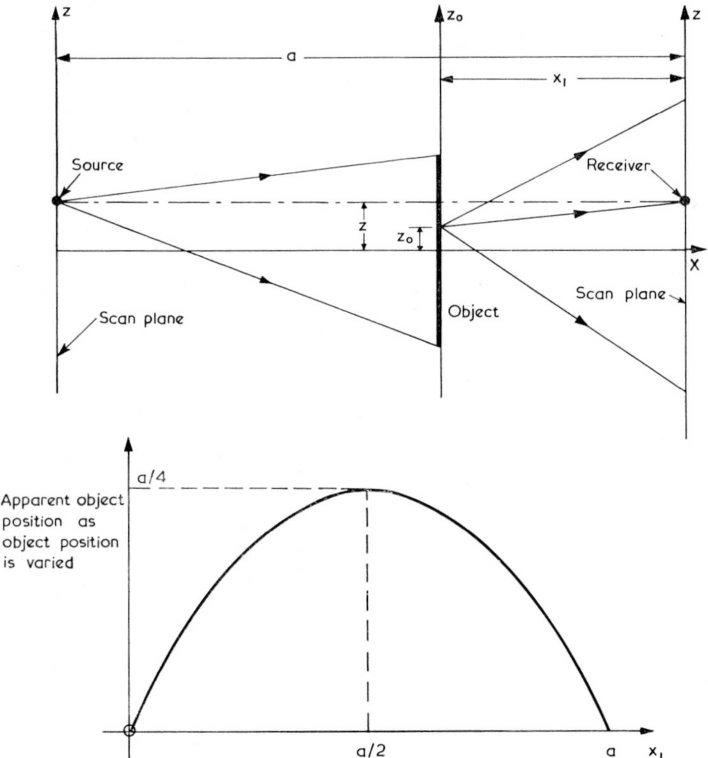

Fig. A.3. Scanning source and receiver in transmission viewing

appears as

$$\frac{1}{x_2}\,e^{i\left(\omega t - \frac{2\pi}{\lambda}x_2\right)} \cdot e^{-i\frac{\pi}{\lambda x_2}(z_0 - z)^2}$$

Denoting the reflectivity of the object as $f(z_0)$ as before, then
the signal registered by the point receiver is given by:

$$\frac{1}{x_1 x_2}\,e^{i\left\{\omega t - \frac{2\pi}{\lambda}(x_1 + x_2)\right\}} \int_{z_0} f(z_0)\,e^{-i\frac{\pi}{\lambda}\left(\frac{1}{x_1} + \frac{1}{x_2}\right)(z - z_0)^2} \cdot dz_0 \qquad \text{A.2}$$

where the integration is over the plane of the object. Comparing expression A.1 with expression A.2, it is seen that they are similar in form except that the object appears as if it were at a distance $x_1 x_2/(x_1 + x_2)$ from the scanning plane of the receiver; the term outside the integral sign is merely a constant under the assumptions made. Since this distance is smaller than the true distance, the image resolution where this is aperture limited, is increased over that of the plane illuminating beam. Also, it is apparent that receiver and source can be interchanged without any change in the image or its apparent position.

If the object is placed between the source and receiver, as shown in Fig. A3, similar arguments apply. If the distance between source and receiver is fixed at a value "a", then the apparent distance that the object is from the receiver plane is $x_1(1 - x_1/a)$. This relationship is also shown in Fig. A3 and four features are apparent. These are that the longest apparent object distance from the receiver plane occurs when the object is midway between the source and the receiver and has the value $a/4$; two object positions lying symmetrically about this mid-point are imaged in the same plane; of the two images corresponding to these object positions, one is pseudoscopic, i.e. convexities become concavities and vice versa; in the region of the mid-position depth compression occurs, i.e. the object appears to be thinner than it really is. To some extent this last feature might be used to compensate for the depth distortion, due to the wavelength change, in the optical image obtained from an acoustical hologram.

(d) Simulation of a Point Receiver by means of the Focal Point of a Focussed Receiver

The simplest argument is since a focussed transducer, when used as a source, concentrates its radiation into its focal point then by the principle of reciprocity, when a wavefront impinges on the same transducer, only that part of the wavefront which passes through the focal point will give rise to signal output from the transducer.

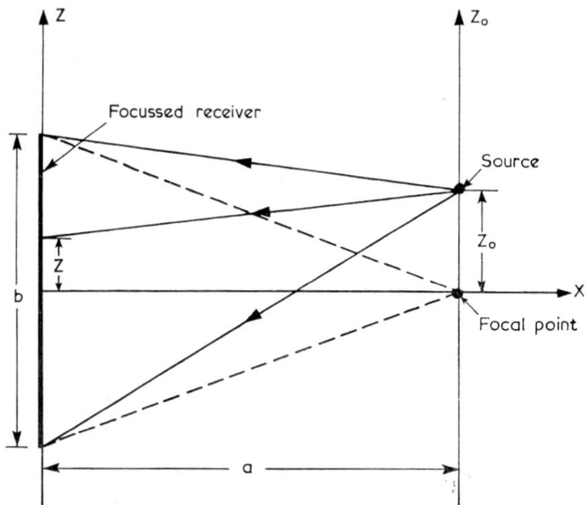

Fig. A.4. Reception of focussed transducer

The following, based on small angle theory, is a more formal argument. When the aperture, b, of the transducer is small compared to its focal length, a, then the transducer can be represented as a plane transducer in which the waveform impinging on it is first multiplied by a phase function, $\exp\left(i\dfrac{\pi}{\lambda a}z^2\right)$, before being integrated over the aperture. As in Fig. A4, let the plane of the transducer be represented by z and the parallel plane through the focal point by z_0. Now consider a point source at z_0 radiating towards the transducer. The wavefront arriving at point z on the transducer is:

$$\frac{1}{a}e^{i\left(\omega t-\frac{2\pi}{\lambda}a\right)} \cdot e^{-i\frac{\pi}{\lambda a}(z_0-z)^2}$$

The signal output of the transducer is then

$$\frac{1}{a}e^{i\left(\omega t-\frac{2\pi}{\lambda}a\right)}\int_{-b/2}^{b/2} e^{-i\frac{\pi}{\lambda a}(z_0-z)^2} \cdot e^{i\frac{\pi}{\lambda a}z^2}\, dz$$

35

which equals

$$\frac{1}{a} e^{i\left(\omega t - \frac{2\pi}{\lambda} a - \frac{\pi}{\lambda a} z_0{}^2\right)} \cdot \frac{\sin\left(\dfrac{\pi b}{\lambda a} z_0\right)}{\dfrac{\pi}{\lambda a} z_0} \qquad \text{A.3}$$

The second part of expression A.3 approximates to a delta function sited at zero value of z_0, i.e. the focal point of the transducer. Thus the transducer will respond only to sources which are close to its focal point. A measure of the closeness can be obtained by finding the smallest value of z_0 which makes expression A.3 zero, i.e.

$$z_0 = \pm \frac{\lambda a}{b} \qquad \text{A.4}$$

This is also a convenient measure of the resolving power of the transducer, as b/a approximates to its cone angle. The criteria usually used is 1·22 times expression A.4, and corresponds to a transducer with circular symmetry.

(e) Cylindrical Scanning

This is scanning in cylindrical geometry such as might arise when a cylindrical object is viewed. In such a case the receiver would be moved slowly parallel to the axis of the cylinder whilst the cylinder itself would be rotated on its axis producing overall a fine pitch helical scan of the cylinder.

It will be assumed for clarity, that the cylinder is self luminous. In cylindrical geometry, let a typical point, r_0, θ, z_0, be radiating to a point receiver situated at R_0, α, z, where R_0 is greater than r_0. The distance, r, between the radiating point and the receiver is given by:

$$r^2 = R_0{}^2 + r_0{}^2 - 2R_0 r_0 \cos(\theta - \alpha) + (z - z_0)^2$$

Since the initial assumptions are that the angles are small, expanding the cosine gives:

$$r^2 \sim (R_0 - r_0)^2 + R_0 r_0(\theta - \alpha)^2 + (z - z_0)^2$$

Assuming $(R_0 - r_0)$ is much larger than the other terms:

$$r \sim (R_0 - r_0) + \frac{R_0 r_0 (\theta - \alpha)^2}{2(R_0 - r_0)} + \frac{(z - z_0)^2}{2(R_0 - r_0)}$$

Thus the signal produced by the point receiver due to radiation from the cylinder is given by:

$$\int \frac{f(r_0 \theta z_0)}{r} e^{\, i \left(\omega t - \frac{2\pi}{\lambda} r \right)} . \, r_0 d\theta dz_0$$

$$\sim \frac{1}{(R_0 - r_0)} e^{\, i \left\{ \omega t - \frac{2\pi}{\lambda} (R_0 - r_0) \right\}}$$

$$\int f(r_0 \theta z_0) e^{\, -i \frac{\pi}{\lambda} \frac{R_0 r_0 (\theta - \alpha)^2}{(R_0 - r_0)}} . \, e^{\, -i \frac{\pi}{\lambda} \frac{(z - z_0)^2}{(R_0 - r_0)}} . \, r_0 d\theta dz_0 \qquad \text{A.5}$$

Comparing expression A.5 with expression A.1, it is seen that the former will produce an astigmatic image. Whilst lines which are parallel to the axis of the cylinder appear as if they were in a plane at a distance $(R_0 - r_0)$, similar to a cartesian scan, lines which are parallel to the circumference of the cylinder appear as if they were in a plane at a distance $r_0 \left(1 - \frac{r_0}{R_0} \right)$. Where r_0 is very much smaller than R_0, this means that the apparent positions of the latter lines are independent of the distance between them and the receiver and merely depend upon the radius at which they lie.

LIST OF SYMBOLS

a distance between source and receiver

b aperture of focussed receiver

f reflectivity of object

m photographic reduction

r radial co-ordinate

r_0, θ, z_0 co-ordinates on surface of self luminous cylinder

R_0, α, z co-ordinates on cylindrical scanning plane

t time

x cartesian co-ordinate

x_0 distance between object and holographic plane

x_1 distance between receiver and object

x_2 distance between source and object

z plane of hologram

z_0 plane of object

A amplitude of acoustical ray

A amplitude of object wavefront

B constant

B amplitude of reference beam wavefront

α inclination of object ray

α angular aperture of object

γ phase of reference beam wavefront

ϕ phase of acoustical ray

ϕ phase of object wavefront

θ inclination of reference beam

ω acoustical frequency

ω_1 optical frequency

λ acoustical wavelength

λ_1 optical wavelength

INDEX